A PEEK AT FLOWERS

DOUG BRADLEY

Published in 2026 by The Rosen Publishing Group, Inc.
2544 Clinton Street, Buffalo, NY 14224

First Edition

Editor: Greg Roza
Book Design: Leslie Taylor

Photo Credits: Cover, p. 1 DrShutter/Shutterstock.com; pp. 4, 6, 8, 10, 12, 14, 16, 18, 20 Series art (background image) LedyX/Shutterstock.com; p. 5 brgfx/Shutterstock.com; p. 7 (clockwise, upper left) ra3rn/Shutterstock.com, (upper right) alias612/Shutterstock.com, (lower left) Nella/Shutterstock.com, (lower right) Studio 888/Shutterstock.com; p. 9 (clockwise, upper left) Janny2/Shutterstock.com, (upper right) Spring_summer/Shutterstock,com, (lower left) Anastasiia Malinich/Shutterstock.com, (lower right) icemanphotos/Shutterstock.com; p. 11 HTWE/Shutterstock.com; p. 13 Marie C Fields/Shutterstock.com; p. 15 Lima_84/Shutterstock.com; p. 17 (left) Nature's clicks/Shutterstock.com, (right) Vadym Lavra; p. 19 (tulip farm) Lukas Gojda/Shutterstock.com, (inset) Olga_Ionina/Shutterstock.com; p. 21 narikan/Shutterstock.com.

Library of Congress Cataloging-in-Publication Data

Names: Bradley, Doug, 1971- author.
Title: A peek at flowers / Doug Bradley.
Description: [Buffalo] : PowerKids Press, [2026] | Series: A peek at plants
  | Includes bibliographical references and index.
Identifiers: LCCN 2024058918 (print) | LCCN 2024058919 (ebook) | ISBN
  9781499451955 (library binding) | ISBN 9781499451948 (paperback) | ISBN
  9781499451962 (ebook)
Subjects: LCSH: Flowers–Juvenile literature. | Plants–Juvenile
  literature.
Classification: LCC QK49 .B73 2026 (print) | LCC QK49 (ebook) | DDC
  582.13–dc23/eng/20250107
LC record available at https://lccn.loc.gov/2024058918
LC ebook record available at https://lccn.loc.gov/2024058919

Manufactured in the United States of America

CPSIA Compliance Information: Batch #CSPK26. For Further Information contact Rosen Publishing at 1-800-237-9932.

Find us on 

CONTENTS

All the Parts

Plants have many different parts! The stem is thick and keeps the plant standing up. Leaves are flat and make food for the plant. Roots grow under the ground and take in water. Many plants also have fruits and flowers. Let's learn about flowers!

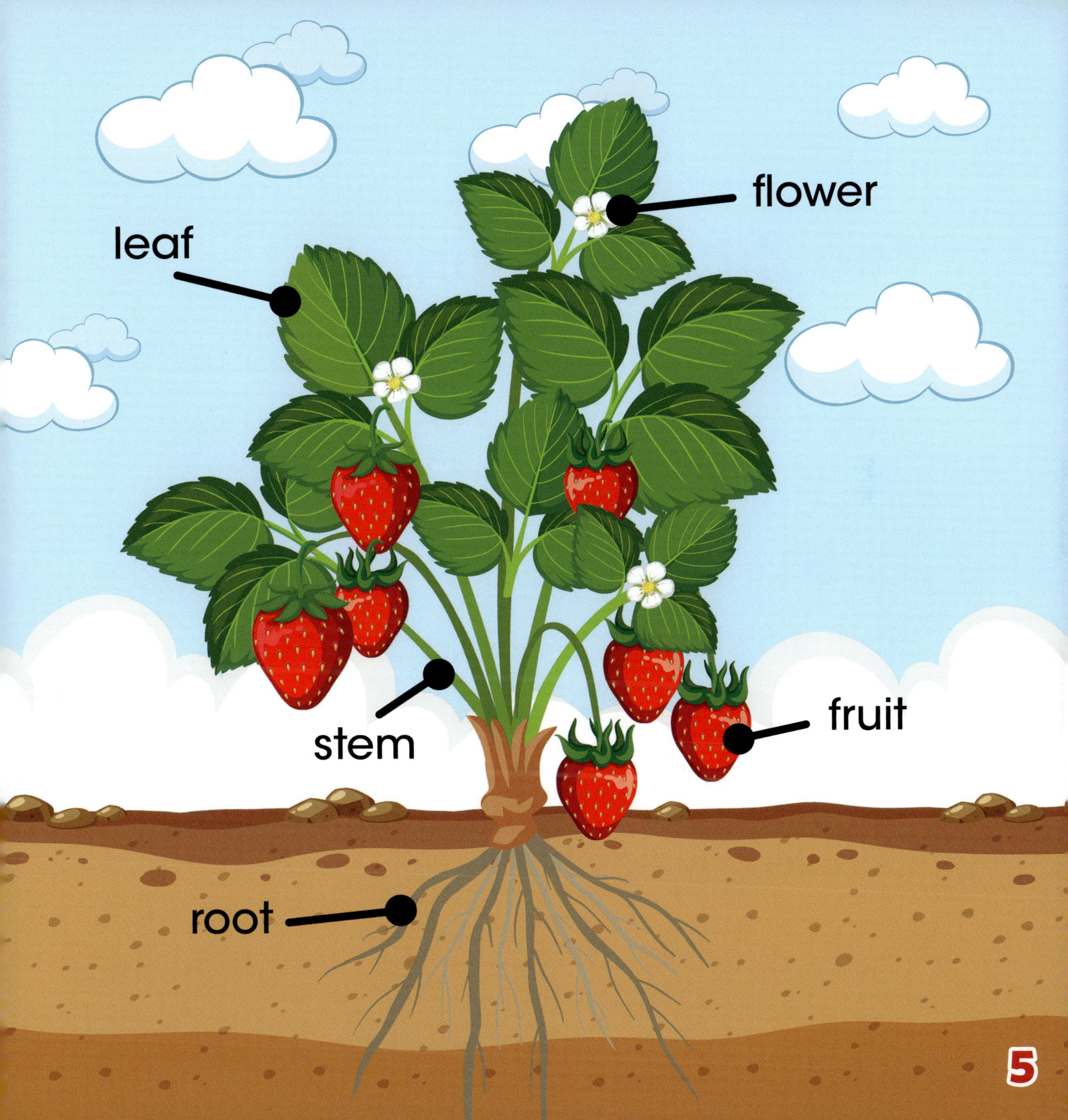

flower
leaf
stem
fruit
root
5

Beautiful Blossoms!

Flowers are also called blossoms. They come in many shapes, sizes, and colors. You can likely name some kinds of flowers—such as roses and daisies. Some blossoms, such as baby's breath, are small. Others, like sunflowers, are big. What's your favorite flower?

roses

daisies

baby's breath

sunflower

A World Full of Flowers

Scientists have found about 300,000 different kinds of flowering plants! These plants grow naturally all over the world. Some flowering plants even grow in **Antarctica**! Scientists think there's likely even more flowering plants out there to find. Have you ever seen any of these flowers where you live?

From Bud to Bloom

Flowers start as a bud on a plant's stem or **branch**. The bud is a small mass circled by tiny leaves called sepals. The bud grows larger, and the sepals spread out. Soon the **petals** come out and the flower blooms, or unfolds.

bud
sepal
petals
branch

Let's Look Inside

There are more special plant parts inside a flower! These parts let plants reproduce, or make more plants. One of these parts makes pollen. Pollen looks like fine dust. Another part has a space where pollen can be used to make seeds.

Helpful Critters

Colorful blossoms draw in bugs and birds. These animals like a **liquid** that flowers make called nectar. Pollen sticks to the bugs and birds. When they fly to a new flower, the pollen drops off. This is how bugs and birds help plants reproduce.

Bye-Bye, Flower

Seeds begin to grow inside the flower. Most flowers last for only a few weeks or months. The petals soon **wither** away. The place where the seeds are grows larger. In time, this plant part becomes fruit! Fruit keeps plant seeds safe.

orange blossoms

oranges

Flower Farms

People have raised flowering plants for a long time. Most early farms were for growing food, but many food plants also make flowers. Today, there are large flower farms all over the world. Some, like this tulip farm in Holland, just grow one type of flowering plant.

tulip

Plant a Garden

You can plan a garden and grow your own flowers! There are so many different kinds to choose from. Some flowering plants, such as tulips, grow every spring. Other flowering plants, such as sunflowers, need to be planted every year. What would you plant in your flower garden?

GLOSSARY

Antarctica: The large landmass that surrounds the South Pole of Earth and is covered by snow and ice.

branch: A smaller part of a plant's stem that grows off the stem and often grows leaves, fruits, and flowers.

liquid: Something that flows like water and takes the shape of the container it is in.

petal: One of the colorful leaves that make up a blossom.

scientist: A person who observes the world and thinks about how things work.

wither: To dry up and grow smaller.

FOR MORE INFORMATION

BOOKS

Grider, Clayton. *How Do Seeds Grow into Gardens? A Hands-on Book About Gardening.* Franklin, TN: Flowerpot Press, 2025.

Silvora Collections. *Plant Parts and Their Purpose: A Budding Botanist's Activity Book.* Independently published, 2024.

WEBSITES

Ducksters: Flowering Plants
www.ducksters.com/science/biology/flowering_plants.php
Learn a lot more about plants and flowers at the Ducksters website.

What Is Pollination?
www.youtube.com/watch?v=UkjOeD47sTs
This short and humorous video uses animation and photographs to explain what pollination is and how it works.

INDEX